An Introduction to
ECOLOGY
AND
POPULATION
BIOLOGY

Nature is but a name for an effect,
Whose cause is God.

William Cowper

In Nature's infinite book of secrecy
A little I can read.

William Shakespeare